Bibliografische Information der Deutschen Nationalbibliothek:

Die Deutsche Bibliothek verzeichnet diese Publikation in der Deutschen National-bibliografie; detaillierte bibliografische Daten sind im Internet über http://dnb.d-nb.de/ abrufbar.

Impressum:

Copyright © 2015 GRIN Verlag, Open Publishing GmbH
Druck und Bindung: Books on Demand GmbH, Norderstedt Germany
ISBN: 9783668354791

Dieses Buch bei GRIN:

http://www.grin.com/de/e-book/342543/technical-and-economic-evaluation-of-biogas-production-from-cow-dung

Nnaemeka Odionye

Technical and economic evaluation of biogas production from cow dung

GRIN Verlag

TECHNICAL AND ECONOMIC EVALUATION OF BIOGAS PRODUCTION FROM COW DUNG

NNAEMEKA ODIONYE

Nomenclature

BoD	Biological Oxygen Demand	H_2	Hydrogen
CDM	Clean Development Mechanism	H_2O	Water
CCD	Central Composite Design	**HRT**	Hydraulic Retention Time
CER	Certified Emission Reduction	**IPCC**	International Panel on Climate Change
CH₄	Methane	**Kg**	Kilogram
Co₂	Carbon dioxide	**Kg/m³**	Kilogram Per Meter Cube
GJ/t	Giga Joules per tonne	**WtE**	Waste to energy

Abstract

A small scale floating drum biodigester was constructed with low cost, recyclable materials and the biogas generated was optimised with an aim to get the best possible combination of parameters to yield optimum biogas through the aid of a Mini-tab software version 17.0. The volume of biogas produced was 0.098 cubic meters, while the optimisation temperature was 28.4 degrees Celsius, being the highest temperature recorded and pH was 7.1. The initial capital invested took 17 years to breakeven.

Table of Contents

1.0 Introduction

Waste to Energy refers to the generation of energy from waste materials such as sewage waste, landfill waste, agricultural vestiges, biomass, and wood-related wastes, which when subjected to various processes can be used in generating energy, particularly electricity or biogas. Humanity over the millennia has utilised environmental resources as the foundation for its developmental drive. The underside to the development drive of the human race is the enormous amount of waste generated by industrialisation and urbanisation which seems inevitably on the rise (Graziano and Matteo, 2010).

Environment can be described as the totality of our surroundings. It consists of both the natural sphere, which encompasses land, air, water, fauna, and flora, as well as the anthropogenic sphere consisting of cities, settlements and other human-induced structures. Furthermore, Odafivwotu and Godwin, (2015) also defined environment as a complex of physical, chemical, and biotic factors such as climate, soil, and living things that act upon an organism or an ecological community and ultimately determines its form and survival . The Bruntland commission in 1987 succinctly defined Sustainable development as *"the development that meets the needs of the present without compromising the ability of future generations to meet their own needs"*, this definition contains within it two key concepts; that of "needs" and "limitations". This is captured in the definition stated above where the present generation has to concentrate in meeting its needs whilst remaining within set limitations regarding use and consumption of environmental resources in order to ensure future generations will have access to environmental resources needed to meet their developmental needs.

1.1 Types of waste

Agricultural Residue: These are residues gotten from harvesting of crops e.g. shafts from grains and animal droppings that have high organic content (Diji, 2013). They are mostly left as useless by farmers and in some cases are burned to make way for the next planting season. When these wastes are burned they emit carbon dioxide which has dire effects on the environment.

Forestry Residues: These are wood fuels produced from existing lumbering operations in established forestry such as wood chips, forestry trimmings, sawdust and bark (Diji, 2013).They could also appear naturally from the falling of trees due to natural causes such as rain, erosion and strong winds.

Animal Waste: These are waste gotten from livestock as well as humans. Livestock waste can be a daunting task to dispose especially for large scale dairy/poultry farms. Findings by Edirin and Nosa, (2012) revealed that in 1985 the number of cattle, sheep, goats, horses and pigs as well as poultry in Nigeria was about 166 million this also amounts to about 227,500 tonnes of waste. Furthermore, Dayo (2008) and Illoeje (2004) estimated that Nigeria produces about 61 million tonnes of animal waste. These could also be further estimated to be about 2.93 X 10^9 KWh and 7.85 X 10^{11} respectively (Sambo, 2005)

Municipal Waste: These are wastes gotten from household, industrial and commercial sources. This waste can be raw, i.e. unsegregated or segregated (glass, metal paper etc.). (Diji, 2013).

1.2 Biogas

Biogas is a combustible mixture of gases gotten from anaerobic decomposition of organic compounds and constituting mainly of methane, carbon dioxide and traces of water, ammonia, hydrogen sulphide and nitrogen (Meshach, 2010). Biogas has a heat content of 7.5GJ/t and density of 1.15kg/m^3. Biogas composition is largely dependent on the organic material/compound being decomposed. When the fatty content of the organic material is high, the methane yield would be equally high. However, when it contains more of glucose, cellulose or semi-cellulosic materials the methane yield is mostly lower (Peter, 2009). Biogas is naturally found in swamps, lakes, tundra, oceans, bowel of ruminant animals and termites. Although, temperature increment could facilitate the emission of biogas from its natural sequesters into the atmosphere.

1.3 Methane

Methane is a major constituent of biogas which is colourless and odourless. It is the first and simplest member of the paraffin hydrocarbon group, it has a boiling point of -162°C and density of approximately 0.75. Methane gas has a global warming potential 23 times that of carbon dioxide. This makes it a major greenhouse gas, as it contributes about 20% of total greenhouse effect caused by anthropogenic activities.

Table 1.1: Sources of methane and content

Substrate	Methane content (%)
Cow manure	65
Poultry manure	60
Pig manure	67
Farm yard	55
Straw	59
Grass	70
Leaves	58
House hold waste	50
Algae	63
Water hyacinths	52

(Source: Ludwig Sasse 1988)

1.4 Uses of Biogas

Biogas like many other fuels has several uses both domestically and industrially as a source of energy in the form of heat or other secondary forms of energy such as electricity etc. The possible uses of biogas are elaborated below:

I. **Cooking:** Biogas is mostly used as cooking gas especially in developing countries where low income earners find it difficult to cope with persistent increase in the price of conventional cooking fuels due to market volatility of the sources of conventional fuels. Biogas as a cooking gas is being encouraged in developing countries as a means of discouraging the unsustainable use of wood as cooking fuel. In addition, biogas is a clean fuel and poses no hazardous treat when used indoors compared to wood fuels. Biogas is about 60 percent more efficient than wood, this in turn frees more time for the cooks especially women to indulge in other duties of choice.

II. **Lighting:** Biogas can be used as fuel for lighting in rural areas and in areas with poor quality of electricity. There are specialised household gauze mantle lamps consuming 0.07 to $0.14m^3$ of gas (Meshach, 2010).

III. **Refrigeration** Biogas can also be utilized for refrigeration, especially on automatic thermo-siphon machines operating on ammonia and water. Also, Refrigerators that run on kerosene flame could be converted to run on biogas (Meshach, 2010).

IV. **Biogas as Mechanical Fuel**

Biogas can be used to run mechanical engines such as industrial machines, vehicles, auto-rickshaws, vehicles etc. Biogas is compactable with four stroke diesel and ignition spark engines. Nonetheless, biogas has some impurities such as hydrogen sulphide and water molecules that could damage engines. Hence, for biogas to be utilised, it has to be purified by passing the gas through a wire gauze. Meshach (2010) also discovered that biogas is being used as a mechanical fuel in Nepal to power irrigation pumps.

V. **Electricity Generation**

Meshach, (2010) resolved that the use of biogas for electricity is a much more efficient use compared to using the resource for lighting from an energy standpoint. Biogas to electricity is mostly utilised in rural electrification and to power farms where cow dungs are used as anaerobic substrates. The gas consumption is about $0.75m^3$ kw/hr with which 25-40 watt lamps can be lighted for one hour.

VI.   Heating

Biogas can be used for space heating in cold areas. This could be achieved with the help of a heat converter/exchanger that takes the warm air from the biogas combustion chamber to the homes or area of need.

1.5 Advantages of Biogas Utilisation

Biogas has diverse advantages both at the microeconomic and macroeconomic level, impacting the lives of individuals, communities within reach for good. Below are various advantages of biogas utilisation on a small/large scale.

1. **Quality Sanitation and Hygiene:** Biogas could be seen as a by-product of a quality/integrated waste management system. Furthermore, anaerobic digestion involves the use of waste, which when left unattended to could result to the development of pathogens and diseases which could have serious consequences on the individual or communities harbouring them.

2. **Social Empowerment:** Women are mostly responsible for cooking in most homes in developing countries. In addition, biogas utilisation for cooking is known not to produce sooth and more efficient than conventional kerosene stoves. Furthermore, domestic biogas utilisation makes available time for women of which they could invest in more productive activities such as paid work, education, recreation etc.

3. **Environmental Improvement:** In developing countries wood fuel is mostly used for cooking, this practice has resulted in the decimation of forests reserves that act as carbon sinks and protects the soil against erosion etc. Biogas utilisation has tremendous impact on the environment by preserving forest reserves that could have been cut down for the sake of cooking fuel.

4. **Microeconomic Income Generation:** The use of biogas is not only advantageous to women by freeing up time for women that engage in cooking but also men/women that are involved in the construction of biodigesters. The skill gained in biogas construction could be put into good use by further construction and maintenance of subsequent biogas plants, providing income for the apprentice and worker.

5. **Macroeconomic Value Addition:** Energy gotten from biogas generation would reflect in the economy as value is added in sectors such as agriculture by increasing the marginal productivity of farms by the sludge fertilizer and energy generation. Also, biogas construction skills would create skilled employment for youths which would reflect on the economy.

1.6 Potentials of Waste to Energy in Nigeria

The repressor defined waste as any substance which constitutes a scrap material or an effluent or any unwanted surplus substance from the application of any process. The conversion of garbage into useful materials involves the use of other useful materials. An example to further illustrate this point is the recycling of newspaper which would always require a lot of energy. Adejobi and Olorunnimbe (2012) established in a survey conducted in Lagos metropolis on the composition of household waste revealed as follows: 56% food waste, 12% paper, 10% plastic, 7% glass, 5%metal, 6% textiles and 4% miscellaneous. Food waste being the highest constituent of waste decays and when disposed inappropriately can provide a conducive environment for bacterial growth hence, a haven for germs and diseases of all sorts.

Recycling of biodegradable organic waste is crucial to meeting the requirements of the landfill directives. However Adejobi and Olorunnimbe, (2012) also stated that by utilising land fill waste disposal system we are limiting the potential for reuse and recycling of valuable resources, thereby increasing demand for new resources and generating more greenhouse gases into the atmosphere which apparently is not sustainable. Olokesusi, (1994) research on ring road refuse disposal system in Ibadan discovered that waste disposal facilities are often perceived to have a negative social impact by host communities. Adesina, (1983) described household waste as one of the easiest to monitor and reduce. It is observed that the rate of waste collection and disposal lag behind the rate at which this wastes are generated, this is the reason why waste are littered around cities (Uwadiegwu and Chukwu, 2013). Uchegbu, (1988) noted that big cities like Enugu, Lagos, Kano and Port Harcourt produce an average of 46kg of waste per individual daily. Adejobi and Olorunnimbe, (2012) stated that an average Lagos inhabitant generates 1 kg of waste daily. As living standards rise, the volume of waste also is bound to increase, Rosenbaum, (1974) established that waste production has often been seen unofficially to reflect economic prosperity as wealthier nations produce more waste (Uwadiegwu and Chukwu, 2013)

Table 1.2: Major Nigerian cities and their waste production

City	Population	Tonnage/ month	Density (Kg/m^3)	Kg/capita/day
Lagos	8,029,200	255,556	294	0.63
Kano	3,248,700	156,676	290	0.56
Ibadan	307,840	135,391	330	0.51
Kaduna	1,458,900	114,433	320	0.58
Port-Harcourt	117,825	1,053,900	300	0.60
Makurdi	249,000	24,242	340	0.48
Onitsha	509,500	84,137	310	0.53
Nsukka	100,700	12,000	370	0.44
Abuja	159,900	14,785	280	0.66

(Source: Ogwueleka, 2009)

2.0 METHODOLOGY

2.1 Digester Design and Construction

A floating drum biodigester design was adopted for anaerobic digestion because of its ease of technicality and the efficiency in measuring daily gas production. It would has a feed-in inlet, an outlet that extends downwards for slurry collection and a drain for routine maintenance when necessary. In conformity to conventional floating drum designs, the gas container is incorporated into the digester and protrudes upwards as gas is produced subsequently. A shaft was incorporated to stir the digestate in the digester so as ensure maximum result by maintaining an even digestate mixture.

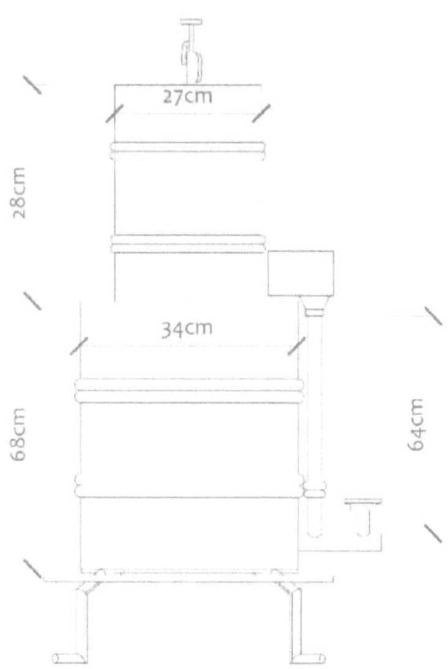

Fig 2.1 Biodigester Design

2.1.1 Construction material

1. Metal container
2. Galvanized pipe
3. Inlet pipe
4. Top bearing
5. Washer
6. Bolt
7. Gas tap
8. Barometer
9. Plumbing tread
10. Gas pipe
11. Gas holder container

2.1.2 Digester Size and Operating Volume

The volume of a digester (V_o) is a function of its substrate volume (Ahmadu, 2009). While the retention time is the interval of time during which the substrate is allowed to ferment within the digester. The retention time is also determined by the temperature and the amount of substrate available (Meshach, 2010). Kossman and Poritz, (1998) stated that a simple design plant should have a retention time of not less than 30 days.

Substrate (S_d) = Biomass (Kg) + Water (L).............................2.1

It is worthy of note that the ratio of water to substrate is 1:1

2.1.3 Total Digester Volume

The total volume (V_t) of the digester should be greater than the operating digester volume (Meshach, 2010). This is to make provision for the biogas and the rise of slurry during fermentation. Ahmadu, 2009 also stated that the operating volume should not exceed 90% of the total digester.

$$V_t = Vo/0.8 \ ....2.2$$

2.1.4 Gas Holder Volume (V_g)

The size of a gas holder is a function of the relative gas generation rate and average gas consumption rate of user (Kossman and Poritz, 1998).

The gas holder size should be able to meet the following requirement

    I.   Cover the peak consumption rate (gc_{max}) for the period of maximum consumption (Meshach, 2010)

        (tc_{max}), $V_g = V_{g2}$ 2.3

II.   To be able to contain the biogas produced even when the gas is not being utilised (t_o)

        (t_o), $V_g = V_{g2}$ 2.4

        From equation 2.3

        $V_{g2} = gc_{max} \ x \ Tc_{max}$ 2.5

        From equation 2.4

        $V_{g2} = G_h \ x \ t_{o \ max}$ 2.6

        Where,

        Gc_{max} = maximum hours of gas consumption (m^3/hr)

        Tc_{max} = time of maximum consumption (hr)

        G_h = hourly gas production (m^3/hr) = G/24hr

        G = Daily gas production (m^3/day)

        T_o = maximum duration when gas is not being utilized (hrs)

The larger value V_{g2} would be the actual size of the gas holder because of its larger size. A safety margin of 10-20% should be incorporated (Ahmadu, 2009).

2.1.5 Force on Gas Holder

The force at which the gas is emitted from the gas holder is a function of pressure and force exerted on the gas holder which could be external.

$$F_g = P_g \times A_g$$

P_g and A_g are the pressure on the gas holder and cross sectional area of gas holder respectively (Meshach, 2010).

2.1.6 Loading of Substrate

The substrate utilised for anaerobic digestion consists of 15kg cow dung. Cow dung was collected from the University of Ibadan dairy farm. The wastes was grinded and mixed to achieve an even mixture with intent of exposing a large surface area for chemical reaction. As earlier indicated the waste was mixed with water in equal proportion in the ratio of 1:1. The waste is then loaded into the biodigester via a funnel into the gas inlet.

2.1.7   Daily Measurement Parameters

pH Value: The pH of the biodigester would be frequently monitored with an aim of detecting the rate of gas production in respect to pH changes. These measurements would be carried out with the help of a pH meter.

Temperature: Chemical reactions are highly dependent on the temperature of reaction, anaerobic reaction is no exception. Hence it is therefore imperative to measure the temperature variance as biogas is produced within the digester.

Daily Gas production: The rate of biogas production would be measured per day to ascertain peak production. Biogas produced per day would be a measure of the daily difference in rise of the gas holder and would be measured as follows (Meshach, 2010).

$G_d = r\pi^2 h_d$

2.2 Response Surface Methodology

Response surface methodology (RSM), which is a collection of mathematical and statistical techniques based on a fit of a polynomial equation to the experimental data. It is applied when a response or set of responses of interest are influenced by several variables. When the dataset does not present curvature, first-order models such as factorial designs can be used. Below are steps in the RSM optimisation technique.

1. The selection of independent variables of major effects on the system through screening studies and the delimitation of the experimental region, according to the objective of the study.
2. The choice of the experimental design and carrying out the experiments according to the selected experimental matrix.
3. The mathematic–statistical treatment of the obtained experimental data through the fit of a polynomial function.
4. The evaluation of the model's fitness.
5. The verification of the necessity and possibility of performing a displacement in direction to the optimal region.
6. Obtaining the optimum values for each studied variable.

$$y = \beta_o + \sum_{i=1}^{k} \beta_i x_i + \sum_{i=1}^{k} \beta_{ii} x_i^2 + \sum_{1 \le i \le j}^{k} \beta_{ij} x_i x_j + \varepsilon,$$

2.3 Financial Analysis

Cost is represented as the price of biodigester and all accessories used in in the construction, including labour. Nonetheless, other forms of intangible cost and benefits are excluded.

Cost of digester $(B_c) = \sum(P_{dc} + P_P + P_a + P_l + P_{gh} + P_m)$

Where P_{dc} is the price of digester container, P_p is the price of PVC pipes, P_a is the price of digester accessories, P_c is the price of construction while P_{gh} and P_m are the price of gas holder and other miscellaneous cost such as transportation etc.

2.3.1 Estimation of Net Present Value of biogas

Cash flow $(C_i) = $ $(extrinsic\ Cost\ of\ biogas + \ldots\ldots + n)$

"C_i" is the benefit for the entire lifespan of the digester and is equivalent to the monetary equivalent of the biogas produced all through the digester duration, "n" is subsequent years which the biodigester would be in use. The steel drums were used for digester construction. They are known to have an average lifespan of 25 years depending on handling, so therefore it is also assumed to be the life span of the biodigester.

$$PV = \sum_i^T \frac{C_i}{(1+r)^i}$$

$NPV = PV\ (C_i) - PV\ (B_c)$

A profitable business decision is one where the future benefit exceeds the cost incurred in the course of the investment.

Therefore,

$NPV \geq 0$

3.0 Discussion

Anaerobic digestion of cow dung was successfully carried out in the constructed digester. The process was carried out outdoors, which is ideal for most domestic biodigesters, which are mostly located outside the home and are often used as organic waste bins. Several parameters such as temperature, pH and daily gas production were monitored with intent on determining factors that enhance biogas production within the digester. Below, table 3.1 represents several dimensions of biodigester and accessories.

Table 3.1: Dimensions of biodigester

Item	Total volume (L)	Operating volume	Length (cm)	Diameter (cm)
Digester tank	36	281	68	34
Gas holder	12	-	28	27
Gas pipe	-	-	200	-
Inlet pipe	-	-	64	2
Outlet pipe	-	-	10	1

3.1 Technical Performance and Production of Biogas

Biogas production was mostly on a daily basis with exception of the first two days where no gas was produced. The average daily production was 0.003 m^3 and the maximum amount of daily gas production was 0.0069 m^3 *see* fig 3.1. The temperature of the solution varied between 24.6-28.4°C, this could be attributed to the alternation of daily ambience temperature, *see* fig 3.2. The pH of the digester also varied as the anaerobic process proceeded in successive days anticipated from a neutral solution to a slightly acidic ph. Although the change in daily pH never exceeded +0.3 or -0.3 which is reflected in fig 3.3

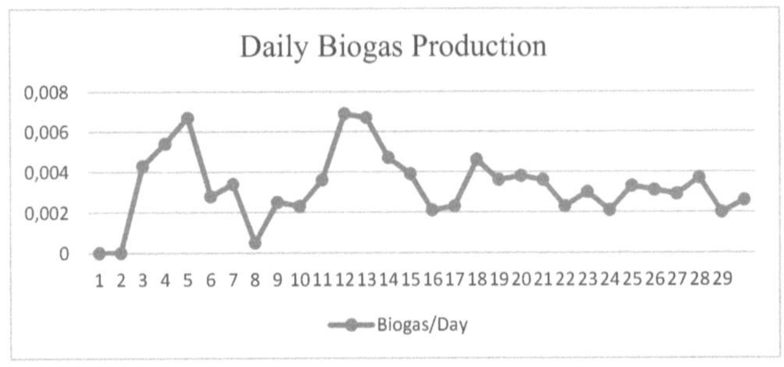

Figure 3.1: Graphical representation of daily gas production

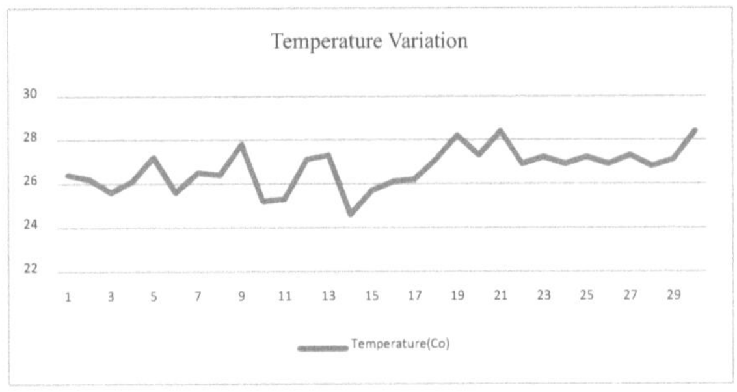

Figure 3.2: Daily temperature value

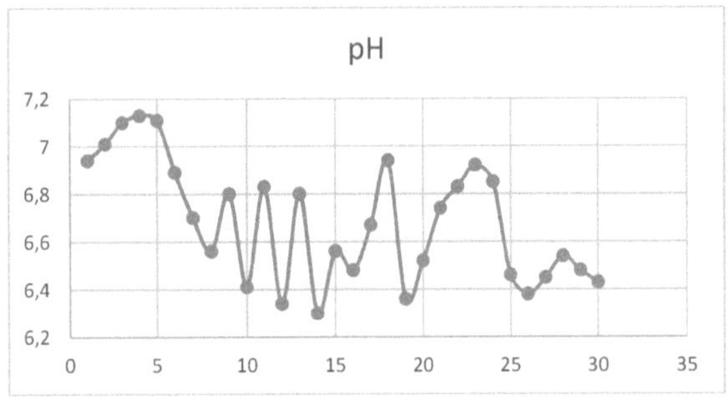

Figure 3.3: Daily pH value

3.2   Results of Temperature and pH Optimisation with RSM

The central composite design was applied to investigate the effect of temperature and pH on the volume of biogas produced from the process so as to determine the combination of both variables that would produce the optimal amount of biogas. Below fig 3.4 shows the surface plot of the process.

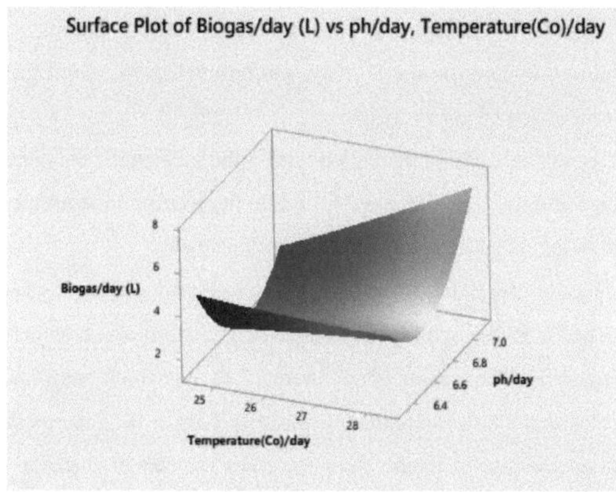

Figure 3.4: Surface plot of biogas/day vs. daily pH and Temperature

3.3 Net Present Value

The total cost of construction of biodigester encompassing materials and construction cost was N 7,100 while the cost of biogas generated from the anaerobic digestion is equivalent to N 1140 annually (Cash flow) sequel to the current natural gas price. Below table 3.4 is a breakdown of economic analysis.

Table 3.2: Breakdown of financial analysis

Item	Cost (Naira)
Total cost of Digester	7,100
Annual Cost of biogas	1,140
Interest Rate	14%
Present Value (PV)	7835.07
Net present Value (NPV)	735
IRR	38%
Breakeven time	17 years

4.0 Summary

The biodigester (floating drum) effectively met the construction requirement which was to be able to measure the daily gas production. The entire digestion process was uncontrolled and proceeded as it would have in its natural state.

1. The initial/early volume of gas generated was due to ongoing anaerobic digestion of cow dung before it was loaded into the digester. Subsequently production became smooth with minimal gas production of $0.0005m^3$ on the 8th Day, gas production increased furthermore and in the third quarter production began to decline.

2. The temperature of the digester fluctuated within the mesophilic temperature range; with the minimum temperature recorded being 24.6 while the maximum temperature recorded is 28.4 this is accrued to the rain season temperature of Ibadan city.

3. The pH of the biodigester changed from neutral to a slightly acidic pH which is expected as the process proceeded from the hydrolysis phase to acidogenesis and acetogensis phase.

4. The RSM technique optimisation temperature is $28.4^{\circ}c$ which happened to be the optimal temperature recorded in the experiment. This is consistent with existing theories which stipulates that an increase in temperature increases the rate of chemical reaction. While the optimisation pH was 7.13 although the pH cannot be controlled as it is determined by the process as the reaction proceeds to subsequent phases.

5. Economically biogas generation from digester proved viable from a rational financial stand point as a positive Net Present Value was recorded as a result of the utilisation of low cost and readily available materials. However, such projects would be more viable when other tangible and intangible factors are considered such as Certified Emission Reduction (CER) cost and cost of reduction in health hazard accrued to the use of alternative fossil fuels etc.

4.1 Challenges

In the course of this experiments several challenges were encountered which might have invariably affected the outcome of the experiments. Listed below are likely challenges faced:

1. The continuous but slow steering of the digestate in the biodigester to get an even mixture. Thereby increasing surface area for reaction.

2. The temperature of the system varied within the day, as the morning temperature deferred from the afternoon temperature which is usually higher as a result of heat from the Sun.

3. The inability to measure the carbon to nitrogen ratio of digestate.

4. The variables utilised for optimisation were few.

4.2 Recommendation

1. Robust campaign to encourage the use of biodigester in homes and farms as a waste control mechanism.
2. Increased government funding for renewable energy projects especially in biogas utilisation; as construction materials are locally available.
3. Local large and small scale farms adopt the use of biodigester proportionate to their waste generation to maximize profit and waste reduction.
4. It should be mandatory that government and private waste management corporation setup a waste recycling plant so as to gain value addition from national waste resources.
5. Existing landfills should be closed and new landfilled constructed with intent on collecting the biogas generated and averting further deterioration of the environment by the leakage of toxins into the underground water.
6. Robust adoption and deployment of biodigesters in homes and societies as a means of climate change mitigation.
7. Effective training of unemployed youth in the area of biodigester construction as a means of empowerment and job creation in the society.
8. The temperature of biodigester should be taken in the morning, afternoon and night and average temperature should be taken to get a more precise temperature reading.
9. To optimise the technical performance of the biodigester more variables should be added such as carbon to nitrogen ratio.
10. Further incorporation of factors such as environmental impact and value of statistical life when calculating the NPV of biogas project.

4.3 Conclusion

Anaerobic biogas generation from biodigester is a proven and reliable technology that should be encouraged to be utilised in place of unsustainable use of firewood especially in developing countries so as to improve our environment as well as local value addition. However, the long breakeven time for capital investment could be reduced by the use of cheap and readily available materials in the construction of biodigesters. Finally, government intervention in the form of tax waivers and free import duty for imported construction materials in the case of large and hi-tech gasifiers would be most welcomed.

References

Adejobi .O and Olorunimbe .R (2012). *Challenges of Waste Management and Climate Change in Nigeria: Lagos State metropolis experience.* African scientific research, 7(1) 346-362

Adesina H.O (1983). *Urban environment and epidemic in E.A Adeniyi, (Eds):* Proceeding of National Conference on Development and Environment. Ibadan. NISER, pp 234-256

Dayo, F.B (2008). *Clean Energy Investment in Nigeria, the Domestic Context.* A Case study for international institute for sustainable development (IISN). Available on http://www.iisd.org

Diji, C.J. (2013). *Electricity production from biomass in Nigeria: Options prospects and challenges.* International journal of Engineering and applied Sciences. [online] 3(4) p. 84-98 Available on www.eaas-journal.org

Edirin, B. and Ogie .N (2012). *A comprehensive review of biomass resources and biofuel production potential in Nigeria.* Research journal in engineering and applied science [online] 1(3) 149-155 available on www.emergingresources.org

Graziano .A and Matteo .F (2010). *The Environmental Kuznets curve in the municipal solid waste sector.* Higher education and research on mobility regulation and the economics of local service.

Kossman .W and Poritz .U (1998). *Biogas digester Vol 4 information and advisory service on appropriate technology-* country reports 13-16

Iloeje .O.C (2004). *Overview of Renewable Energy in Nigeria, Opportunities for rural development and development of renewable energy masterplan.* Paper presented at the renewable energy conference "Energetic solutions" Abuja/Calabar 21-26 Nov. 2004

Ludwig Sasse (1988). Biogas plants Deutsches Zentrum fur Entwicklungstechnologien available online http://www.susana.org/_resources/documents/default/2-1799-biogasplants.pdf

Meshach I.A (2010). *Comprehensive study of biogas production from cow dung, chicken droppings and Cambopgon Citratus as alternative source of energy in Nigeria.* A thesis presented to the department of water resources and environmental engineering, Ahmadu Bello University Zaria, Nigeria

Odafivwotu   Ohno   and   Godwin   Ekpo   (2015).   *Environmental   Perception*
http://www.nou.edu.ng/uploads/noun_ocl/pdf/edited_pdfs/esm%20403%20envt%20perception
%20unedited.pdf

Ogwueleka T. Ch (2009). *Municipal solid waste characteristics and management in Nigeria.*
Iran.J.Environ.Health.Sci.Eng   Vol.6   No.3   PP.   173-180   [online]   available   on
http://www.bioline.org.br/pdf?se09026

Olokesusi .F (1994). *Impact of the ring road waste disposal facility in Ibadan.* The Environmentalist
Vol 7 pp 55-60

Peter .J (2009). *Biogas-green energy process, design, energy supply, environment* [online] available on
http://lemvigbiogas.com

Rosenbaum W.A (1974). *The politics of Environmental Concern.* New York, Praeger printers

Sambo A.S (2005). *Renewable Energy for Rural Development: The Nigerian Perspective.* ISESCO
Science and Technology Vision, Vol 1 12-22

Uwadiegwu B.O and Chukwu K.E (2013). *Strategies for effective Urban and Solid Waste Management
in Nigeria.*European Scientific Journal Vol.9, No 8

BEI GRIN MACHT SICH IHR WISSEN BEZAHLT

- Wir veröffentlichen Ihre Hausarbeit,
  Bachelor- und Masterarbeit

- Ihr eigenes eBook und Buch -
  weltweit in allen wichtigen Shops

- Verdienen Sie an jedem Verkauf

Jetzt bei www.GRIN.com hochladen
und kostenlos publizieren